Hamshack Raspberry Pi

A Beginner's Guide to The Raspberry Pi for Amateur Radio Activities

Jeremy Stephens

© 2017

© Copyright 2017 by Jeremy Stephens
All rights reserved.

This document is geared toward providing exact and reliable information in regard to the topic and issue covered. The publication is sold with the idea that the publisher is not required to render accounting, officially permitted, or otherwise, qualified services. If advice is necessary, legal or professional, a practiced individual in the profession should be ordered.

- From a Declaration of Principles which was accepted and approved equally by a Committee of the American Bar Association and a Committee of Publishers and Associations.

In no way is it legal to reproduce, duplicate, or transmit any part of this document in either electronic means or in printed format. Recording of this publication is strictly prohibited and any storage of this document is not allowed unless with written permission from the publisher. All rights reserved.

The information provided herein is stated to be truthful and consistent, in that any liability, in terms of inattention or otherwise, by any usage or abuse of any policies, processes, or directions contained within is the solitary and utter responsibility of the recipient reader. Under no circumstances will any legal responsibility or blame be held against the publisher for any reparation, damages, or monetary loss due to the information herein, either directly or indirectly.

Respective authors own all copyrights not held by the publisher.

The information herein is offered for informational purposes solely, and is universal as so. The presentation of the information is without contract or any type of guarantee assurance.

The trademarks that are used are without any consent, and the publication of the trademark is without permission or backing by the trademark owner. All trademarks and brands within this book are for clarifying purposes only and are the owned by the owners themselves, not affiliated with this document.

TABLE OF CONTENTS

INTRODUCTION 5
INTRODUCING RASPBERRY PI 6
   A Quick Ride Through History 7
   The Hardware 9
GETTING STARTED WITH PI 13
   The Installation 13
   User Management 21
   Accessing Pi Remotely 24
   Wireless Connectivity 34
   Installing Samba 38
SETTING UP A RASPBERRY PI PACKET RADIO ... 40
   Setting up the System 41
   Installing and Configuring AX.25 45
   Configuring the Packet Node 50
   Starting the Node at Boot 56
   Playing Zork Over the Air 59
HAM RADIO PROJECTS FOR THE RPI 63
   Digital Modulation with Fldigi 63
   Ham Radio World Clock 67
   SDR and GNU Radio 68
   Amateur Satellite Tracking 73
   XLog Logging 75
   Graphical Transceiver Control Program 77
   Morse Code Virtual Radio 79
CONCLUSION 96

Introduction

Whether you have already received your Ham license, or plan on receiving one, choosing to read this book is definitely a smart move.

Everything cool that you have ever heard about radio stations such as communicating in Morse code, tracking satellites, or even playing a simple game over the air, this book will teach you in no time.

And the best part? You don't have to spend thousands or even hundreds of dollars to do so. A cheap Raspberry Pi computer and some other inexpensive tools, is all you need to start your amateur radio journey.

And if you think that just because you haven't been properly introduced to the tiny computer called Raspberry Pi is a reason to skip reading this book, then you couldn't be more wrong.

From the moment you buy your Pi, to the actual playing with your hamshack RPi, this book will guide you through the whole process of learning to install, configure, and operate with Raspberry Pi in order to enjoy the coolest hobby in the world.

Sounds like a great deal? Now let's play with the airwaves, shall we?

Introducing Raspberry Pi

Raspberry Pi (pronounced *raspberry pie)* is a tiny computer that has the size of a credit card. Seriously, the Raspberry Pi foundation used an actual credit card to be their template for the design of the printed circuit board (PCB) of this small computer.

Raspberry Pi, or *the Pi,* or *RasPi*, or *RPi*, or whatever nickname you like the most, believe it or not is an actual computer. Its board features the typical hardware that is found in the desktop computers (such as RAM, processor, etc.), which means that this credit card-sized computer also enables you to edit documents, play audio and video, play games, do some coding, etc.

Of course, its tiny size is not able to provide as much power as a normal desktop PC, however, its cheap price of approximately $35 makes up for everything. Actually, the main idea behind the birth of such a computer was to teach kids and adults about the basics of computer science with a minimal investment. It's a lot more convenient to break a Raspberry Pi and replace it with a new one, than learn about the software science with an actual desktop computer and then spend a fortune to replace or fix what you've broken.

Raspberry Pi is as yummy as its sounds. With this tiny computer you can actually move past the 'visible', surface-level software and dive deeply into its 'black box' – the internals that most people are unaware of. Starting the software education with a Raspberry Pi has proven to be the easiest way for people to adopt highly appreciated and super valuable software and hardware engineering skills.

But Raspberry Pi isn't only for these 'academics'. What's even more amazing about this credit card-like computer is that its fan base is super versatile. There are many DIYers and hackers that make Raspberry Pi an essential part of their computer experiments.

A Quick Ride Through History

For those that are simply casually observing the development of the computer technology, it may seem that Raspberry Pi is brand new. And it'd seem like it, since most blogs and websites treat this computer that way. However, the truth is that Raspberry Pi has been around for years. In fact, the creators of the Pi – Rob Mullins, Eben Upton, Alan Mycroft, and Jack Lang – first came with the idea of making a tiny computer in 2006. This came as a result of their observance that the available cheap computers at that time (such as the Amiga or the

Spectrum) had a rather negative effect on the programming education, as they were slowing down the ability to learn the software science significantly. And since laptops and desktop computers cost hundreds, some even thousands of dollars, kids back then really couldn't afford to mess with the main family computer that way. So it is only understandable just how much a cheap learning platform was needed.

Trying to realize their idea, the Pi creators played around with various microcontrollers, PCBs, and breadboards, but it wasn't until 2008 that their concept became a reality thanks to the newer technology and cheaper chips. The new powerful tools helped them create a platform that was not only able to teach command-line programming but also supported media. Joined by David Braben and Pete Lomas, the original creators formed the now-popular Raspberry Pi Foundation, and only three years later, in 2011, the first Raspberry Pi hit the market.

With the rise of technology many different models of Raspberry Pi have been created, each of them offering better features than its ancestors.

The Hardware

If you have already purchased your Raspberry Pi, you may have already noticed how 'naked' the device actually comes. Its price may be cheap, about $35, but there are some hidden costs involved. The Pi may be a computer on its own for sure, however, you need a couple of other things to make it work as it should.

A 5V Power Supply. For a device that is USB-powered, everyone can agree that the Raspberry Pi, no matter which flavor you have, is pretty hungry for power. It draws around 600 to 700 mA. And while it can be powered from the USB port – which is rated at around 500 mA, it is important that you use an actual powered adapter. You can use a modern smartphone charger since most of them supply 700 mA at 5V, however, check to see if yours do at the bottom of the charger.
If you don't have a good quality charger, once other devices such as a camera module or a simple Wi-FI dongle are connected, they will draw even more power from the Pi and it may become unstable.

On the other hand, if the power supply doesn't deliver solid 5V, and maybe provides too much power, the board can easily get fried.
Great third-party vendors who can provide your Pi with adequate power are:

- AdaFruit
- MoodMyPi
- SparkFun

SD Card. Its small and its super powerful, but the Raspberry Pi does not have the capability to store data onboard. That is why, and SD (Secure Digital) Card – a removable storage device – is essential. Here is what you need to look for when shopping for an SD card for your Raspberry Pi:

- A Standard SD Card. Your Pi supports a standard SD card, not a Mini or a Micro SD, however, you can use an adapter to convert these if you happen to have a Micro or Mini.

- A trustworthy brand. Avoid the cheap choices and choose good quality SD card such as Kingston, SanDisk or Transcend.

- At least 4 GB of capacity

- Class 4 or higher. The class is important because it indicates how quickly the card can actually write and read data. For instance, an SD card of class 4 can read and write MB per second. Class 6 is faster with 6 MB per second, class 10 can read 10 MB per second, etc.

Keep in mind that the Standard SD card comes in SDHC (Secure Digital High Capacity) and SDXC (Secure Digital eXtended Capacity). Make sure to check your Pi's compatibility and see which one is best for your device. The SDHC goes up to 32 GB, and the SDXC up to 2T.

Powered USB Hub. An USB hub is a compact device that can host a couple of USB devices. For instance, if you have the low-cost Raspberry Pi, you may have only one or two ports. Plug your mouse and keyboard and you have no way to plug anything else. That is why an USB hub is one of the main things that Raspberry Pi owners buy right after purchasing their tiny computer.

Ethernet Cable. If you want to connect your Raspberry Pi to the Internet (and of course you do), then, besides the obvious Internet connectivity, you will need an Ethernet cable to do so. You plug one end of the Ethernet cable in your Pi and the other in your Wireless router, cable modem, or whatever device you have for connecting to the internet.

Monitor. Even if you plan on using your Raspberry PI heedlessly and remotely, chances are that sooner rather than later, you'll want to plug your computer into a monitor or a TV.

Cables. To plug your Raspberry Pi into a monitor or TV, you will need cables. For this purpose, you will probably need an HDMI cable or a composite video cable.

In case you will be using analog video, you will also need a 3.5 mm stereo audio cable whether you want to get a sound after connecting it to your TV or monitor, or plan on connecting it to external speakers.

USB Mouse and Keyboard. As a computer, the Raspberry Pi also requires a keyboard and a mouse. You can use both wired and wireless, however, know that many users have reported some issues with the wireless keyboards and mice, as most Pi models require that you unplug and plug back in, when the Pi reboots.

Getting Started with Pi

Now that you know what you need to start your Raspberry journey, the next step is to actually get started with this amazing tiny computer. Before we begin setting up the radio and learn about different radio program for your Ham Shack home, it is essential that we first learn how to actually set everything up in order to be able to let your DJ skills run wild.

In this chapter you will learn everything about installing and getting started with your brand new Raspberry Pi. I promise that when you get to the next chapter, you will no longer feel intimidated by the raw computer board you bought the other day. So let's get started, shall we?

The Installation

Now, it is time to actually install an operating system on your Raspberry Pi. Unlike traditional computers, the OS on the Raspberry Pi is not installed on a hard drive (since the Pi doesn't have one), but on the SD card. There are many ways in which you can install an OS on your Pi, however, this is the best and most convenient one.

NOOBS and BerryBoot

NOOBS or New Out of the Box Setup is the official installation method created by the Raspberry Pi Foundation, with the purpose of simplifying things for beginners and adding multiple support tools for different operating systems on a single card. It is the BerryBoot precedes NOOBS and it actually supports more operating systems, has the ability to actually install them on a USB stick, as well as back up, clone, and restore OS.

Here is what you need to do in order to install NOOBS and BerryBoot:

- Go to http://www.berryterminal.com/doku.php/berryboot and download BerryBoot, or go to http://www.raspberrypi.org/downloads to download NOOBS.
- Depending on which OS you use, format your SD card as FAT32.
- If you have Windows, then it is suggested that you use the SD Association's SD Formatter. Go to https://www.sdcard.org/downloads/formatter_4/ to access it.
  If you are a Linux user, stick to GParted.
- Now it is time to extract the downloaded archive on the SD card.

Installing Raspbian with RAW Images

This is the original method used for installing an operating system on an SD card. The RAW images have all the necessary binary data, including the Master Boot Record, as well as all the partitions.

The biggest advantage of this method is that it supports every OS, and there is not much that can go wrong, however, the disadvantage lies in the fact that it is not that flexible. It is somewhat hard to boot multiple system on the same card. Besides, the image file has a static partition table which means that you will have a lot of free, but unused space.

Writing Images with Windows

- Go to https://launchpad.net/win32-image-writer and download the Win32 Disk.

- Select the right image file.

- In the drop-down menu, find and select your SD card.

- Click 'Write' and eject the card.

Writing Images with Linux

- First, make sure that the SD card is not mounted.
- Replace the `mmcblk0` with the right device in order to write to the disk. To do this, you have to run this command:
  `# dd if=file.img of=/dev/mmcblk0 bs=4M`

- It is important that you run this command: `# sync`, in order to flush the filesystem buffers before you remove the SD card.

Booting

Now that you have installed Raspbian and have the image written, it is time to perform the initial setup by booting your Pi for the very first time. The best way to do so is by using a keyboard, a monitor, and a mouse.

Insert the card and connect the monitor, keyboard, and mouse. When asked for a username, type `pi`, and when asked for a password, type `raspberry`. Finally, type `startx` and load the familiar UI.

After that simply connect the Ethernet cable to the Raspberry Pi.

Expanding the Filesystem

If you have Raspbian, you can do this automatically by choosing *Expand Filesystem* found in the `raspi-config` script. There is also a way for you to manually launch it if you choose to run `sudo raspi-config`. However, if you are not provided with an automated way to do it, you can actually expand the filesystem on the Raspberry Pi.

1. Launch the fdisk by typing:
   `# fdisk /dev/mmcblk0`
2. Press *p* and then *enter*. The display should read something like:

```
Command (m for help): p
Disk /dev/mmcblk0: 29.7 GiB,
31918653440 bytes, 62341120
sectors
Units: sectors of 1 * 512 = 512
bytes
Sector size (logical/physical):
512 bytes / 512 bytes
I/O size (minimum/optimal): 512
bytes / 512 bytes
Disklabel type: dos
Disk identifier: 0x417ee54b

Device          Boot     Start
End   Blocks  Id System
```

```
/dev/mmcblk0p1              2048
186367     92160   c W95 FAT32 (LBA)
/dev/mmcblk0p2            186368
3667967  1740800   5 Extended
/dev/mmcblk0p5            188416
3667967  1739776   83 Linux
```

3. The card has an extended partition P2 that has a logical side, a partition p5. You need to remove P2 so that P5 can be deleted automatically.

   ```
   Command (m for help): d
   Partition number (1,2,5, default 5): 2

   Partition 2 has been deleted.
   ```

4. Now you need to create a new extended partition in place of the one you have just deleted. The first sector of that partition has to match the sector of the deleted partition. After that, simply accept the recommended last sector so that it can use up the free space automatically:

   ```
   Command (m for help): n

   Partition type:
   ```

```
   p   primary (1 primary, 0
extended, 3 free)
   e   extended
Select (default p): e
Partition number (2-4, default
2):
First sector (186368-62341119,
default 186368):
Last sector, +sectors or
+size{K,M,G,T,P}       (186368-
62341119, default 62341119):

Created a new partition 2 of
type 'Extended' and of size 29.7
GiB.
```

5. Create a new logical partition instead of the deleted one and also make sure that this one matches the old one's start sector:

```
Command (m for help): n

Partition type:
   p   primary (1 primary, 1
extended, 2 free)
   l   logical (numbered from 5)
Select (default p): l

Adding logical partition 5
```

```
First sector (188416-62341119,
default 188416):
Last sector, +sectors or
+size{K,M,G,T,P} (188416-
62341119, default 62341119):

Created a new partition 5 of
type 'Linux' and of size 29.7
GiB.
```

6. Display the partition table and make sure that the end sectors are the only things that are different:

```
Command (m for help): p
Disk /dev/mmcblk0: 29.7 GiB,
31918653440 bytes, 62341120
sectors
Units: sectors of 1 * 512 = 512
bytes
Sector size (logical/physical):
512 bytes / 512 bytes
I/O size (minimum/optimal): 512
bytes / 512 bytes
Disklabel type: dos
Disk identifier: 0x417ee54b

          Device
Boot       Start          End
Blocks        Id System
```

```
/dev/mmcblk0p1            2048
186367        92160        c
W95     FAT32 (LBA)
/dev/mmcblk0p2           186368
62341119  31077376     5
Extended
/dev/mmcblk0p5           188416
62341119  31076352    83
Linux
```

7. Write the partition table to disk:

   ```
   Command (m for help): w
   ```

8. Restart your Raspberry Pi:

   ```
   # reboot
   ```

9. Now login to resize the filesystem:

   ```
   # resize2fs /dev/mmcblk0
   ```

User Management

After installing your OS, the next step you should take is to set up an actual user account. As you have probably noticed, the Raspbian has already a default user called *pi* and a default password *raspberry*. So why change them? Well, the username and password is widely known and a system that runs with these default setting is really not secure at all. If you want to avoid any security issues in the future, I suggest you change your password.

The installation doesn't have a root password set, so the user should use sudo to run the commands that way. It is important to have a root password set because it will allow for the administrative tasks to be performed in the root shell. Here is how you can standardize the install:

1. Log in as a *pi* user.

2. Change the default password with this command:

    ```
    # passwd
    ```

3. Enter a root shell:

    ```
    # sudo -i
    ```

4. Run passwd one more time to change the password. Since you are in a root shell this will automatically change the root password.

5. Choose a new user name and add it with this command (I have used John as a new user name in this example):

   ```
   # adduser john
   ```

6. If you want your user to perform certain tasks, it has to be a part of certain groups. To ensure that, type the command below. Again, replace john with your user name of choice:

   ```
   # usermod -a -G
   adm,dialout,cdrom,sudo,audio,video,plugdev,games,users,netdev,input,spi,gpio john
   ```

If you want to boot your desktop as a new user, you can actually do this without the `raspi-config`. Here is how to do it:

1. Enable the LightDM service:

   ```
   # update-rc.d lightdmd enable 2
   ```

2. In `/etc/lightdm/lightdm.conf`, set the `autologin-user` variable, as it is required. Know that commenting with a # will provide you with a login window every time that you boot your Pi, which is way

more secure, and especially beneficial if you have more than one user.

If you want to accomplish the reverse, simply disable the LightDM.

Accessing Pi Remotely

After installing your OS, you can start developing with a USB keyboard, USB mouse, and an HDMI screen. However, if you don't have these accessories, all is not lost. You can still access your Pi remotely. There are a couple of ways to do this.

SSH Remote Server

SSH is a great remote server that can be used for most of the remote administration processes. And since Pi is a great 'headless' server, this remote server can be installed by default in the Raspbian. In Linux, you can connect to SSH easily with the `ssh` command:

```
ssh user@IP:port
```

If you are using Windows, the SSH connection can be made with PuTTY, that you can download from http://www.chiark.greenend.org.uk/~sgtatham/putty/.

Securing the SSH

In case you want to access the Pi from outside the local network, then you should be super careful and do everything you can to protect it from unwanted bots and hackers. Most bots tend to connect to the port 22 and then brute force the password, and even if the password is secure, this can still file the logs with multiple failed attempts to establish a connection.

The simplest way to protect yourself is by changing the port number and disabling the option for root login over the SSH. To do this, edit the `/etc/ssh/sshd_config` and set up these options:

```
Port 1286 #Any unused port number is fine
PermitRootLogin no
```

But this won't stop someone from scanning your open ports. The ideal thing is to disable any password login and use only keys for this purpose. To do this, you need to generate two keys – public and private. Send the public key to a remote server, and keep the private key, well, private, and use it instead of a password for login.

If using Windows, you can generate the keys with the PuTTY:

- Copy the public key here `~/.ssh/authorized_keys`, and save the private key on your PC.

- Change the `authorized_keys` permissions, if you want to protect the key:

    `# chmod 600 ~/.ssh/authorized_keys`

- In PuTTY, click 'Save Private Key'.

In Linux, you can generate the keys this way:

- By typing `# ssh-keygen -t rsa`

- Now, accept recommended save location and enter your passphrase. The content from `id_rsa.pb` should be copied to `~/.ssh/authorized_keys`, but that can be easily done with:

    `# ssh-copy-id user@IP -p PORTNUMBER`

- Once you see that you can now connect to your Pi with the private key, go back to

`/etc/ssh/sshd_config` to disable the password logins:

`PasswordAuthentication no`

- The best thing is that you can actually add multiple public keys. For instance, if you want to allow someone else to access your Pi, you can simply ask them for their public key and add the key to your `authorized_keys` file.\

- There is one more thing to do, and that is to install `fail2ban`. This will ban all of the malicious IPs and fight various attacks:

`a# apt-get install fail2ban`

Now, open the configuration file in order to edit the Sudo nano `/etc/fail2ban/jail.local`, and paste this:

```
# SSH
# 3 failed retry: Ban for 10
minutes
[ssh]
enabled = true
port = ssh
filter = sshd
```

```
action = iptables[name=SSH,
port=ssh, protocol=tcp]
mail-whois-
lines[name=%(__name__)s,
dest=%(destemail)s,
logpath=%(logpath)s]
logpath = /var/log/auth.log
maxretry = 3
bantime = 600
ignoreip = 192.168.0.0/10

[ssh-ddos]
enabled = true
port = ssh
filter = sshd-ddos
action = iptables[name=SSH,
port=ssh, protocol=tcp]
logpath = /var/log/auth.log
maxretry = 10
ignoreip = 192.168.0.0/10
```

- Now restart the `fail2ban`;

    ```
    sudo         /etc/init.d/fail2ban
    restart
    ```

Transferring Files

If you thought that SSH was only great for allowing us to execute remote commands, you were wrong. SSH also gives us the chance to transfer files.

If you want to perform a simple file transfer, this command will help you do that:

```
$ scp /path/to/src user@IP:/path/to/dst
```

However, if you are looking for a bit more powerful command, then this is the one:

```
$ rsync /path/to/src user@IP:/path/to/dst -av
```

If you are a Windows user, you can do this through WinSCP, which can be found at http://winscp.net. Its GUI uses PuTTy for private keys, and also allows you to transfer files between your PC and Pi.

X11 Forwarding

Great, you can now execute commands and transfer files remotely. But the other thing you need to have in mind is the GUI. The X Windowing System can use a remote display in mind, and you can use SSH to forward GUI programs to the local X server remotely.

If you use Linux, you can launch SSH as before, only with the addition of the `-Y` option. When your SSH session is ready, launch `leafpad`.

If you are a Windows user, you will need to have an X server running. Here are some options for that:

- Cygwin/X that can be found at http://x.cygwin.com/.
- Xming, found at http://sourceforge.net/projects/xming/.
- VcXsrv that is available here: http://sourceforge.net/projects/vcxsrv/.

Maintaining the Remote Session with Screen

Although *screen* and *ssh* are not exactly related, they make a great combo. Think about what would happen if your connection dropped during a SSH session. Obviously, everything would be lost. That is why *screen* is so important. With *screen*, you can easily re-attach and detach as needed. For instance, if your connection drops, you can re-attach to the screen session and pick up where you left off. And since the Raspbian doesn't have the screen in its default setting, you need to install it with this command:

```
# apt-get install screen
```

After the installation, you can launch it by running:

```
# screen
```

To get started, you need to add some bindings. You can do this with CTRL + and then another key. For instance, if you want to detach the screen session, you can do that by CTRL + and then *d*. Once you detach, you can go back by running:

```
# screen -x
```

The Reverse SSH

If your Pi is behind a firewall and doesn't have access to port forwarding, then you cannot access it directly. But, in a case where the computer that you want to connect to is not behind a firewall and is with an active SSH server, then using a reverse SSH is the best choice.

Start SSH with this command:

```
$ ssh -N -R 2222:localhost:22 user@IP
```

Here, -N means that you don't want to execute other commands, and -R 2222:localhost:2 means that the remote computer will tunnel the port 22 to port 2222.

After that, you can connect to the port 2222 which will then redirect you to port 22 :

```
$ ssh user@localhost -p 2222
```

VNC and Virtual Display

You can also use VNC to access the Pi remotely. VNC comes in two different flavors. The first one *x11vnc* allows that the display that the Pi is attached to is mirrored remotely. The other one is called *virtual display*. Virtual display is not shown in the display, and is useful for those situations when you need to launch multiple VNC sessions that are independent. But, you cannot connect to VNC without a VNC client. A perfect example is *TightVNC*.

So, the only way you can actually use the display is if you have the x11vnc installed, and that can be done by running this:

```
# apt-get install x11vnc
```

However, to use the virtual display you will need to have TightVNC installed, which can be done by running this:

```
# apt-get install tightvncserver
```

When you run the vncserver the first time you will notice a couple of procedures. You need to pay attention to what the output is because there you will find the display that the server is actually running on.

If you want to stop running VNC you can do it by using the `-kill` option. This will kill the server with the display `:1`:

```
$ vncserver -kill :1
```

Synergy

There is also an option of sharing your mouse and keyboard across different systems. For instance, you can have a display that is connected to your Pi, and you can actually run it by the mouse and keyboard that are connected on your PC.

Here is how you can do this:

- Download a synergy server on your PC. You can find it at http://synergy-foss.org. Run it and follow the setup's steps.

- Install it on Pi with this command:

  ```
  # apt-get install synergy
  ```

- Launch Synergy from the Pi's desktop with this command:

  ```
  # synergyc IP
  ```

Wireless Connectivity

It is only obvious that you would want to establish an internet connection on your Raspberry Pi. There are a couple of ways to do so, but since you are probably most interested in setting up a wireless connection, in this section I will talk about how you can get your Pi online wirelessly.

Pixel Desktop

There is a GUI provided that can help you set up the wireless connectivity in Raspbian with PIXEL desktop. If you take a look at the right hand of the Pi's menu bar, you will see the network icon. If your Wi-Fi dongle is plugged in, after clicking on the icon, a list of available wireless networks will appear. Click the network that you wish to connect to. If the network is not secured the connection will be established without additional steps. If it is secured, a dialogue box will appear where you will have to enter the network key. Click 'ok' and wait for the connection establishment.

If there are no available networks found, a message saying "No Aps found – scanning...".

Command Line

If you have no access to Pi's graphical user interface and cannot set up the wireless connectivity that way. Have no access to a screen, or a wired network, then choosing to establish a wireless connection through the command line is the best.

1. First of all, you need to scan for Wi-Fi networks. That can be done with the command: `sudo iwlist wlan0 scan`. This will give you a list of all the available wireless network, as well as some other useful related information. Here is what you should look out for:

- 'ESSID:"testing" - that is the name of the wireless network.

- IE: IEEE 802.11i/WPA2 Version 1' – that is the authentication that has been used. You will also need to know the network's password, which is mainly found at the bottom of the router. The examples below have the ESSID – `testing`, and a password – `testingPassword`.

2. Now, open this configuration file: `wpa-supplicant` in nano:
   `sudo nano /etc/wpa_supplicant/wpa_supplicant.conf`

   Go to the bottom and add paste this:

   ```
   network={
        ssid="testing"
        psk="testingPassword"
   }
   ```

   To generate the encrypted password, you can use `wpa_passphrase`. Using the examples from above, you can generate it with `wpa_passphrase "testing" "testingPassword"`. You will get this output:

   ```
   network={
          ssid="testing"
          #psk="testingPassword"

   psk=131e1e221f6e06e3911a2d11ff2fac9182665c004de85300f9cac208a6a80531
      }
   ```

Keep in mind that the plain version is commented out and you can delete that line from the `wpa_supplicant` at the end, for the sake of security.

This tool requires a password that has at least 8-63 characters. If the passphrase is more complex, you can simply extract the content from a text file and then use it as an input for this `wpa_supplicant` tool. If it is actually kept in the form of plain text inside some file, you can then call: `wpa_passphrase "testing" < file_where_password_is_stored`.
Make sure to delete the `file_where_password_is_stored` at the end, again, for the sake of security.

If your PSK is `wpa_passphrase` encrypted, then you should paste it into `wpa_supplicant.confg`, or simply redirect your output tools with: `wpa_passphrase "testing" "testingPassword" >> /etc/wpa_supplicant/wpa_supplicant.conf`. This means that you'll have to change the root through the `sudo su` command.

Save the file by hitting *Ctrl + X,* then *Y,* and finally *Enter.* Within a few moments you should be able to connect to the network. If the connection cannot be established, you can try restarting the interface by `calling sudo wpa_cli reconfigure.`

Installing Samba

Imagine you have files on your PC that you wish to access on your Raspberry Pi. For that purpose, you will need to have the Samba software installed. But I have no Windows clients in my network, you may think. Well, even if that is the case, you will still benefit a lot from Samba. For instance, you have the Photo Mate RAW – a processing application for Android – can access pictures that are stored on PCs. And if you wish to access files from your Pi on an Android, then Samba is the way to do it.

You can install Samba simply by running:

```
sudo apt-get install samba samba-common-bin
```

Then, add the user *pi* to the `smbdpasswd` file:

```
sudo smbpasswd -a pi
```

In nano, open the `smb.conf` file:

```
sudo nano /etc/samba/smb.conf
```

Make sure to delete the comment in the *security = user* line. Go to the end of the file and then add share definition. You can do that by simply replacing </path/to/dir> with the actual path to the directory where the files are stored. In the case below, we want to access photos:

```
[Photos]
path = </path/to/dir>
available = yes
valid users = @users
read only = no
browsable = yes
public = yes
writable = yes
```

After that, simply save your changes and restart Samba:

```
sudo /etc/init.d/samba restart
```

Setting Up a Raspberry Pi Packet Radio

If you have already received your Ham license, you must be super excited about the amateur radio journey that lies ahead of you. And now that you know how to setup, configure, as well as operate with the amazing Raspberry Pi, then you probably cannot wait for the actual adventure and the chance to experience some great projects with your Hamshack Raspberry Pi. But before we start having fun, there is one huge step that we need to take care of first – setting up the actual radio.

If you think that setting up a radio station is a real hassle, it is probably because it usually is. But that doesn't mean that you have to start your adventure the most expensive way, purchasing high-tech equipment and all sorts of fancy tools. Your Hamshack raspberry pi can be simple, and yet extremely effective in executing all of the most popular ham RPi projects.

Although there are many different ways in which you can turn your tiny computer into an amateur radio station, I chose this simple Raspberry Pi packet radio node as your beginner's starting point, because I believe that is the simplest way for you to build and sending data over the airwaves.

This example turns your Raspberry Pi into a packet radio node, and then runs Zork on it. How amazing is that?

What you will need:

Raspberry Pi (SD card included)
USB to Serial Adapter
Amateur Radio Transceiver and an Antenna
KISS TNC with a Serial Cable
List of Some Pocket Nodes from Your Area

Setting up the System

Although we have already gone into details about how to set up and configure your Raspberry Pi, going through these simple steps again is the perfect way for you to master your berry skills.

The first thing you need to do, obviously, is to install the Raspbian OS. Go back to the beginning of the previous chapter, follow the steps, and install the operating system. Once you do that, boot the device, and do a simple nmap scan of your network to see if you can find your Raspberry Pi:

```
sudo nmap -sP 192.168.1.1-254
```

Let's assume that only one device with a Raspberry Pi address was found

```
Nmap   scan   report   for   raspberrypi
(192.168.1.111)
Host is up (0.0013s latency).
MAC      Address:      B8:27:EB:B6:A6:A3
(Raspberry Pi Foundation)
```

Now, using SSH, connect to the IP with the default credentials. Remember how we said that changing the password is crucial? Well, it is time to do that.

```
pi@raspberrypi:~ $ passwd
Changing password for pi.
(current) UNIX password:
Enter new UNIX password:
Retype new UNIX password:
passwd: password updated successfully
```

After that, update the system:

```
pi@raspberrypi:~    $    sudo    apt-get update
pi@raspberrypi:~    $    sudo    apt-get upgrade
```

Make sure to install all of the updates that are available. Now, let's get rid of your Ethernet cable and set up a wireless connectivity. Sure, if you don't have a Wi-Fi dongle or a card, you can skip this step and continue with your wired Ethernet connection.

Before all, let's make sure that the Wi-Fi card is recognized automatically:

```
pi@raspberrypi:~ $ iwconfig
wlan0           unassociated  Nickname:"<WIFI@REALTEK>"
 Mode:Managed  Frequency=2.412  GHz  Access Point: Not-Associated
 Sensitivity:0/0
 Retry:off   RTS   thr:off   Fragment thr:off
 Power Management:off
 Link Quality:0  Signal level:0  Noise level:0
 Rx invalid nwid:0 Rx invalid crypt:0 Rx invalid frag:0
 Tx excessive retries:0  Invalid misc:0 Missed beacon:0
```

In this case, as you can see, the card is registered as wlano. Let's edit the wpa_supplicant config and actually restart the interface, just like we talked about in the previous chapter:

```
pi@raspberrypi:~ $ sudo nano /etc/wpa_supplicant/wpa_supplicant.conf
```

Just as we previously did, go to the bottom and paste this:

```
network={
 ssid="MYSSID"
 psk="MYWIFIPASSWORD"
}
```

Restart the interface of the network and check to see if it has an IP address:

```
pi@raspberrypi:~ $ sudo ifdown wlan0
pi@raspberrypi:~ $ sudo ifup wlan0
pi@raspberrypi:~ $ ifconfig wlan0

wlan0     Link    encap:Ethernet    HWaddr
74:da:38:5a:6e:e0
 inet                addr:192.168.1.138
Bcast:192.168.1.255
Mask:255.255.255.0
 inet6                             addr:
fe80::26f5:8a4f:824c:b75e/64
Scope:Link
 UP    BROADCAST    RUNNING    MULTICAST
MTU:1500 Metric:1
 RX    packets:24    errors:0    dropped:6
overruns:0 frame:0
 TX    packets:37    errors:0    dropped:0
overruns:0 carrier:0
 collisions:0 txqueuelen:1000
 RX     bytes:2933    (2.8    KiB)    TX
bytes:8033 (7.8 KiB)
```

Installing and Configuring AX.25

Everything that is needed for the packet radio, actually works on top of the protocol of AX25. That is why installing it at this point is essential:

```
pi@raspberrypi:~ $ sudo apt-get
install ax25-tools ax25-node ax25-
apps telnet
```

Now, let's configure the AX.25 port that is actually your packet radio's network interface:

```
pi@raspberrypi:~ $ sudo nano
/etc/ax25/axports
```

The file is supposed two have two example lines that are commented with #. Now add this:

```
0         CALLSIGN-N           9600    255
2         145.030 MHz (1200bps)
```

Let's break down the line from above:

0 – this is the name of the port. It can be either a text or a number

CALLSIGN-N – This is your callsign and it has to have a number after the name.

9600 – This represents the port's speed. Actually, this is not the speed of the radio connection, but the speed between the TNC and the computer. In this case that is 9600 band.

255 – This is the PacLen.

2 – This is the window

145.030 MHz (1200 bps) – This serves as a reminder. You can basically put any text description here that will remind you what this port is.

Moving forward, it is time to set up a netrom (a packet protocol that also runs on top of AX25) port. In order to set up the node, you need to set up the netrom first. If you wish to connect solely to other nodes, then configuring the netrom is not mandatory, however, keep in mind that it will make the process of connecting to remote nodes way much easier.

```
pi@raspberrypi:~ $ sudo nano /etc/ax25/nrports
```

Now, add something like this:

```
netrom CALLSIGN-N #ALIAS 255
145.030 MHz (1200bps)
```

Let's break down once again:

Netrom – The name of the port

CALLSIGN-N – Same as before, the callsign followed by a number

#ALIAS – the alias of the netrom packed code. Think of this as an alternate name under which other people can connect to your node.

255 – PaClen

145.030 MHz (12000 bps) – Just like before this serves as a description of the netrom port.

Let's edit a daemon config file. Delete everything from the file and replace it with the text found below. Be careful, you must edit it in a way that it will match the configured ports:

```
[CALLSIGN-N VIA 0]
NOCALL  *  *  *  *  *  *  L
default  *  *  *  *  *  *  -  root
/usr/sbin/ax25-node ax25-node
```

```
<netrom>
NOCALL  *  *  *  *  *  *  L
default  *  *  *  *  *  *  -  root
/usr/sbin/ax25-node ax25-node
```

Do not forget to update the headings from the section with the actual ones. For instance, the CALLSIGN-N should be your callsign followed by the number you have chosen. VIA 0 should be updated and contain the name of the AX,25 port, etc.

Let's now bring up the interface of the AX25 port:

```
sudo kissattach /dev/ttyUSB0 0
```

/dev/ttyUSB0 0 needs to be replaced with the corresponding serial port for your own KISS TNC. Remember, the 0 here is actually the name of the AX25 file that you configured earlier. By using ifconfig you can now check that the port AX25 is online:

```
pi@raspberrypi:~ $ ifconfig ax0
ax0 Link encap:AMPR AX.25 HWaddr CALLSIGN-N
 UP BROADCAST RUNNING MTU:255 Metric:1
 RX packets:0 errors:0 dropped:0 overruns:0 frame:0
```

```
 TX   packets:0   errors:0   dropped:0
overruns:0 carrier:0
 collisions:0 txqueuelen:10
 RX bytes:0  (0.0  B)  TX  bytes:0  (0.0
B)
```

Now you should have no trouble testing your settings. First, set up a listener on the AX25 port, and then open a new SSH session, and run this command:

```
pi@raspberrypi:~ $ sudo axlisten -c -a
```

After running this program, you will see that the program will go blank. You may think that nothing is going on, however, the program is actually listening to the AX25 port and will display the pockets spotted by your modem. If you stay tuned to a frequency that is active, after some time you will be able to see the packet data displayed.

You can also connect to a local packet station as a different test. Leave the axlisten command running, and you will see the outbound packets, just like any other response, really. This is excellent for debugging.

Assuming that in this example, the radio is tuned to a frequency of 145.030 MHz, let's initiate the connection now:

```
pi@raspberrypi:~ $ axcall 0 W7EUG-10
```

The 0 here represents the axport that we configured before. If you can communicate with the BBS then the axlisten should output something similar to this:

```
Th0:  fm  KE7SAL-9  to  W7EUG-10  ctl
SABM+
0: fm W7EUG-10 to KE7SAL-9 ctl UA-
0: fm W7EUG-10 to KE7SAL-9 ctl I00^
pid=F0(Text) len 40
0000 Lane County ARES/RACES OADN RMS
GatewayM
0: fm KE7SAL-9 to W7EUG-10 ctl RR1v
```

If you want to disconnect at this point, simply type *Bye* and then hit *Enter*.

Configuring the Packet Node

After making sure that the AX.25 is up and running, we can now setup the packet node, in a way that it can startup automatically right after boot, as well as listen for incoming connections with the right prompt, commands, and permissions.

Let's first edit the file in charge for controlling the node's main settings, and that is the node.conf file.

```
pi@raspberrypi:~      $      sudo      nano
/etc/ax25/node.conf
```

Although for most of the file you won't have to change the default settings, the HostName and the NodeId should be altered. Pay attention that you also configure the Nport to the already specified port name.

Here is an example of what that should look like:

```
HostName          KE7SAL-9
NodeId            CARMEL:KE7SAL-9
NrPort            netrom
```

Let's add the node's permission. If you want to, you can make it that way that your user needs a password.

```
Letpi@raspberrypi:~      $      sudo      nano
/etc/ax25/node.perms
```

Read the node.perms man page to see what are the settings and permissions. In this example, this is what the node.perms file looks like:

```
.pe# user type port passwd perms
ke7sal * * PASS 255
```

```
# Default permissions per connection type.
#
* ax25  * * 7
* netrom * * 7
* local * * 7
* ampr  * * 7
* inet  * * 7
* host  * * 7
```

Now, let's configure the broadcasting of netrom. This is important as it will tell the node how to actually process other nodes' broadcasts. Plus, it allows you to configure the settings of your broadcasts.

```
pi@raspberrypi:~ $ sudo nano /etc/ax25/nrbroadcast
```

The main page holds a lot of information. The setup of this example looks like this:

```
# ax25_name min_obs def_qual worst_qual verbose
0 5 192 100 0
```

It is now time to create the netrom port:

pi@raspberrypi:~ $ sudo nrattach netrom

Setting it up as a TCP/IP service will make it a lot easier to connect to the local packet node directly from the shell. Let's define the port and service first though:

```
pi@raspberrypi:~ $ sudo nano /etc/services
```

At the bottom, add this line:

```
ax25-node       4444/tcp        # KE7SAL packet node
```

Here you can actually change the 444 port to any other port that us unused. Let's make the service start automatically right after boot:

```
pi@raspberrypi:~ $ sudo nano /etc/inetd.conf
```

At this at the bottom:

```
ax25-node  stream  tcp  nowait  root /usr/sbin/ax5-node ax25-node
```

Now, you can restart the service:

```
pi@raspberrypi:~ $ sudo service inetd restart
```

All of this enables you to telnet into the local packet node straight from the command line. Also, you can telnet it from any other system, just as long as your Raspberry Pi is on that same network, and the 4444 port is opened to remote the connections.

```
Spi@raspberrypi:~ $ telnet localhost 4444
Trying ::1...
Trying 127.0.0.1...
Connected to localhost.
Escape character is '^]'.

LinuxNode v0.3.2 (KE7SAL-9)

login:
```

Great! Now simply log in with your callsign. You should see something like this:

```
#CARMEL:KE7SAL-9 Welcome to KE7SAL-9 network node

Type ? for a list of commands. help <commandname> gives a description of the named command.

--
```

If you want to get a list of all of the commands, just type '?'. Now, tune the radio to the frequency of the node, and issue a call with AX25. You can do this with your node, by adding:

```
C 0 WALKER
```

Of course you will need to update this command line. C means CONNECT, and it doesn't necessarily have to be changed. 0 represents the AX.25 port that we configured earlier. And WALKER is an alias of a local packet node. You will have to change this with the alias of your own local packet node. You should get something like this:

```
#CARMEL:KE7SAL-9   Trying   WALKER   on
port 0... Type <RETURN> to abort
#CARMEL:KE7SAL-9 Connected to WALKER
on port 0 (Escape: CTRL-T)
```

Now, disconnect:

```
#CARMEL:KE7SAL-9   Trying   WALKER   on
port 0... Type <RETURN> to abort
#CARMEL:KE7SAL-9 Connected to WALKER
on port 0 (Escape: CTRL-T)
bye
#CARMEL:KE7SAL-9    Reconnected     to
KE7SAL-9
```

```
bye
#CARMEL:KE7SAL-9 Goodbye
Connection closed by foreign host.
```

We want our node to listen for other nodes' netrom broadcasts, as well as to ensure that we actually have a server that is listening for other systems' incoming connections. That can be done with two simple lines:

```
pi@raspberrypi:~ $ sudo netromd
pi@raspberrypi:~ $ sudo ax25d
```

And we have successfully configured it.

Starting the Node at Boot

There are a couple of ways to do this, however, we will choose the simplest option. We will simply dump all of the command in one place - /etc/rc.local

```
pi@raspberrypi:~     $     sudo    nano /etc/rc.local
```

Before the exit line, add the text from below:

```
/sbin/modprobe ax25
# Load ax.25 kernel module
```

```
/sbin/modprobe netrom
# Load netrom kernel module
/usr/sbin/kissattach /dev/ttyUSB0 0
# Create the AX.25 interface
/usr/sbin/nrattach netrom
# Create the netrom interface
/usr/sbin/netromd
# Start the netrom service
/usr/sbin/ax25d
# Start ax.25 service and start
listening
```

Now, simply reboot your Raspberry Pi:

```
pi@raspberrypi:~ $ sudo reboot
```

Backing Up the Node List

Usually, the nodes broadcast list every 30 – 60 minutes. Once the node sees a list, it automatically updates its table. After rebooting the node, the list is emptied, and the node becomes more difficult to use. In that case you simply need to wait for more broadcasts. If you want to make things a little bit smoother, back up your node on regular basis, and restore the list at boot.

First, you need to create the restore script:

```
pi@raspberrypi:~    $    sudo    touch
/etc/ax25/nodebackup.sh
```

```
pi@raspberrypi:~ $ sudo chmod u+x /etc/ax25/nodebackup.sh
```

If you create a cronjob you can backup the nodes regularly:

```
pi@raspberrypi:~ $ sudo crontab -e
```

In this example, this line sets it to run every 60 minutes:

```
0 * * * * /usr/sbin/nodesave /etc/ax25/nodebackup.sh
```

Make the file to be executable, and then run it at boot:

```
/etc/ax25/nodebackup.sh
```

You need to wait for the node list to populate first, before checking how this works.

```
pi@raspberrypi:~ $ cat /etc/ax25/nodebackup.sh
```

Now simply reboot. When your Raspberry Pi is online again, log back into the packet node to list the known nodes. If it's not empty, then everything works just fine.

Playing Zork Over the Air

Having a packet node is really amazing. And while the nodes mostly exist for emergency purposes, there are some pretty great uses you can find for your packet radio node. And since everything in this set up will probably cost you around $100, I say that it is a pretty affordable hobby.

If waiting for an emergency or hopping from node to node, mapping them all together gets kind of boring, here is a great use for your brand new radio packet node. With this example, you can actually learn how to play Zork over the air, and if that's not amazing, I really don't know what is.

But in order for you to be able to play Zork on Linux, you will first need an emulator. A great emulator for this purpose is Fortz:

```
pi@raspberrypi:~ $ sudo apt-get install frotz
```

After that, you can simply download Zork directly from infocom:

```
pi@raspberrypi:~ $ mkdir ~/games
pi@raspberrypi:~ $ mkdir ~/games/zork
pi@raspberrypi:~ $ cd ~/games/zork
```

```
pi@raspberrypi:~/games/zork    $    curl
http://www.infocom-
if.org/downloads/zork1.zip > zork.zip
pi@raspberrypi:~/games/zork    $    unzip
zork.zip
```

After your download completes successfully, you can launch Zork locally:

```
pi@raspberrypi:~/games/zork    $    frotz
DATA/ZORK1.DAT
```

Now you have to find a way in which the node users can actually launch the Frotz emulator and play Zork. In order to launch the application the right way, you first have to create a shell script:

```
pi@raspberrypi:~ $ nano zork.sh
```

After you have created the shell script, add this:

```
#!/bin/bash
export HOME="/home/pi"
export TERM="linux"
/usr/games/frotz
/home/pi/games/zork/DATA/ZORK1.DAT
```

Allow the execution:

```
pi@raspberrypi:~ $ chmod u+x zork.sh
```

Normally, when you launch Zork you have to type the last line only. However, when you launch it from the AX25 port, the Frotz will not launch if you do not export the terminal variables. Since we already have the script, the only thing we need to do is to create a new command on the node, so that Zork can be launched:

```
pi@raspberrypi:~ $ sudo nano /etc/ax25/node.conf
```

Go to the bottom of the ExtCmd section, and add the following:

```
ExtCmd  ZOrk  1  pi  /home/pi/zork.sh NULL
```

This will create the command ZOrk. We use capital ZO, so you don't have to use the entire name all the time. Make sure that you have the NULL portion there, otherwise, the note will not execute.

If you type the command "?" in your node, you will get something like this:

```
CARMEL:KE7SAL-9 Commands:
?,  Bye,  Connect,  ECho,  Escape,
Finger, Help, HOst, Info, Links
Mheard, NLinks, Nodes, PIng, Ports,
Routes, Status, TAlk, Telnet, TIme
Users, ZConnect, ZOrk, ZTelnet
```

Now, enter ZOrk, and the game can be launched:

```
ZORK I: The Great Underground Empire
Copyright   (c)    1981,    1982,    1983
Infocom, Inc. All rights reserved.
ZORK  is  a  registered  trademark  of
Infocom, Inc.
Revision 88 / Serial number 840726

West of House
You  are  standing  in  an  open  field
west of a white house, with a boarded
front
door.
There is a small mailbox here.

>
```

And isn't that fun?

I know that this tutorial for setting up a Raspberry Pi packet radio node is somewhat lengthy, but please, do not let the length of the code discourage you. I have really done my best to provide as much information as possible, carefully guiding you through the whole process, step by step. If you stick to the rules from this chapter, you can have your packet radio node up and running in no time.

Ham Radio Projects for the RPi

Once you get your own ham radio up and running, you will notice that there are literally dozens of things you can actually do with it, and not have to use it for emergency purposes only. It may seem overwhelming or even impossible for you to be able to send and track radio signals at this point, but believe me, once you try the amazing projects from this chapter, you will not only see how great of a hobby having a ham radio is, but that it can also be easily mastered.

Although you can try any of these programs first, I really suggest that you follow the order in which they are listed since I have put a lot of thought to arrange them in a way that will help you to quickly master Hamshack. Shall we start?

Digital Modulation with Fldigi

First of all, let me briefly explain what digital modes are and why they are important. Digital modes represent a means of Amateur radio operating with the help of a keyboard. The computer here acts as a modem, or a modulator – demodulator, that allows you to type, as well as to see what someone else types. It is in charge of controlling the transmitter, providing helpful features, as well as changing the modes as needed. In this case, the modes are used on high frequency bands, which is especially the case with the chat modes – those that are used to have a conversation such as Morse – that will let different operators to take part in a single net.

Raspberry Pi can use the USB sound card's audio input in order to decode these digital modes. The most popular and effective program that supports these modes that the ham radio users use today, is Fldigi. This program operates in conjunction with a HF SSB radio transceiver and uses the USB sound card to input from, and sends output to the radio.

The best thing about Fldigi is that it is multi-mode, which basically means that it can operate multiple modes, without having to switch programs. Fldgi includes all of the popular modes:

- THOR
- CW
- PSKR
- Contestia
- Throb
- NAvtext/SitorB
- Hell
- Olivia
- 8PSK
- RTTY

- DomionoX
- MFSK
- MT63
- QPSK
- WEFAX

Here is how you can install Fldgi with the PIXEL Desktop GUI:

1. Open the top left Application Menu from the PIXEL GUI. Select Preferences, and then choose Add/Remove Software

2. Search for Fldgi, and check on the box to select it

3. Find the Apply button and click on it in order to install Fldgi

After installing it, you can launch the program by simply opening the top left Application Menu in the PIXEL GUI and selecting Internet, Fldgi.

Hardware Configuration for Modulation and Demodulation:

- Yaesu Data Cable, CT-39
- CT-39 Data Out cable, 1200 bps – plugged into the FT – 857D

- CT-39 Data Out, connected to the Behringer Xenyx 320 USB's line input
- CT-39 Data In, plugged into the speaker jack of the Raspberry Pi

Hardware Configuration for Radio Control:

- Raspberry Pi P3 Model B
- Yaesu Cat Cable, CT-62
- USB Dual Serial Adapter (Micro Connectors. Inc)
- Yaesu transceiver, FT-857D

Configuring the Ports of the Pi:

- Make sure that the Raspberry Pi P3 Model B is powered on.
- With a powered USB Hub, plug the USB Dual Serial Adapter into the USB port of the Pi
- This is what should appear in the */dev/* directory:
  `dev/ttyUSB0`
  `/dev/ttyUSB1`
- Plug the Yeasu CT-62 Cat Cable into the FT-857D transceiver
- Turn on the FT-857D transceiver
- Connect the Yaesu CT-62 cable to the Serial Dual Adapter RS-232 S1. The RS-232 S1 connector is `/dev/ttyUSB0`.

Ham Radio World Clock

There are a couple of clock programs that you can find, however, I strongly believe that Twclock is by far the best one. This program will not only display the GMT and local time, but it will also display the time from hundreds of cities from all around the world. Besides, it comes with a built-in alarm that will tell you when the time for a station ID comes.
This alarm can actually notify you in a couple of different ways. It can either bling the alarm button, send your call through your sound call in CW, or simply beep the connected speakers.

The installation process is really simple, and it can be done with the CLI:

First, you need to update the Pi's repository index. That can be done with this command:

```
sudo apt-get update
```

Then, you will have to search the repository index and find the twclock program:

```
apt-cache search twclock
```

Finally, you can install the twclock with this command:

```
sudo apt-get install twclock
```

To start the Twclock, open the PIXEL's top left Application Menu, then select Accessories, and Twclock. You will notice that this will open two twclock programs. One of them is for displaying the local time, and the other one is for the GMT time.

If, for any reason, you want to uninstall the twclock, you can do it by typing this command:

```
sudo apt-get uninstall twclock
```

SDR and GNU Radio

Before explaining what GNU Radio is and how to install it, let me just tell you that with Raspberry Pi, you can actually set up the cheapest SDR (Software Defined Radio) receiver. For this purpose, besides the Raspberry PI, you will also need a cheap RTL Dongle, and a USB sound card. And if you think that this is probably shabby, you will be surprised to hear that you can actually receive anywhere from 25 MHz to incredible 1800 MHz. To create a pretty capable scanner, you simply need to attach this to a discone antenna.

Of course, to turn this all into a reality, you will need to run SDR# in the Raspbian.

Now, about the GNU. GNU radio is a free software toolkit that is in charge of processing blocks in order to implement software defined radios. This can be used with some low-cost external hardware to create SDRs, or even in an environment that is simulation-like, without the hardware. This is mostly used among hobbyists who want to support the real-word radios, as well as their Wi-Fi communication research.

In most cases, the applications for GNU Radio are written with the Python programming language, whereas the supplied signal processed path is implemented in the C++ language, and if possible, with some processor-floating extensions.

This particular package comes with the gnuradio – companion, which is a graohical tool that can create signal flow graphs, as well as generate their source code. Many other helpful tools are also included.

The installation method, again, uses the CLI. First of all, you will need to update the repository index of your Raspberry Pi:

```
sudo apt-get update
```

Then, you need to search the index to find gpredict programs. You can do this with this command:

```
apt-cache search gnuradio
```

This is what you should find:

```
pi@raspberrypi:~ $ apt-cache search gnuradio
gnuradio - GNU Radio Software Radio Toolkit
gnuradio-dev - GNU Software Defined Radio toolkit development
gnuradio-doc - GNU Software Defined Radio toolkit documentation
gr-air-modes - Gnuradio Mode-S/ADS-B radio
gr-fcdproplus - Funcube Dongle Pro Plus controller for GNU Radio
gr-osmosdr - Gnuradio blocks from the OsmoSDR project
libair-modes0 - Gnuradio Mode-S/ADS-B radio
libgnuradio-analog3.7.5 - gnuradio analog functions
libgnuradio-atsc3.7.5 - gnuradio atsc functions
libgnuradio-audio3.7.5 - gnuradio audio functions
```

libgnuradio-blocks3.7.5 - gnuradio blocks functions
libgnuradio-channels3.7.5 - gnuradio channels functions
libgnuradio-comedi3.7.5 - gnuradio comedi instrument control functions
libgnuradio-digital3.7.5 - gnuradio digital communications functions
libgnuradio-dtv3.7.5 - gnuradio digital TV signal processing blocks
libgnuradio-fcd3.7.5 - gnuradio FunCube Dongle support
libgnuradio-fcdproplus0 - Funcube Dongle Pro Plus controller for GNU Radio
libgnuradio-fec3.7.5 - gnuradio forward error correction support
libgnuradio-fft3.7.5 - gnuradio fast Fourier transform functions
libgnuradio-filter3.7.5 - gnuradio filter functions
libgnuradio-iqbalance0 - GNU Radio Blind IQ imbalance estimator and correction
libgnuradio-noaa3.7.5 - gnuradio noaa satellite signals functions
libgnuradio-osmosdr0.1.3 - Gnuradio blocks from the OsmoSDR project
libgnuradio-pager3.7.5 - gnuradio pager radio functions

```
libgnuradio-pmt3.7.5 - gnuradio pmt
container library
libgnuradio-qtgui3.7.5 - gnuradio Qt
graphical user interface functions
libgnuradio-runtime3.7.5 - gnuradio
core runtime
libgnuradio-trellis3.7.5 - gnuradio
trellis modulation functions
libgnuradio-uhd3.7.5 - gnuradio
universal hardware driver functions
libgnuradio-video-sdl3.7.5 - gnuradio
video functions
libgnuradio-vocoder3.7.5 - gnuradio
vocoder functions
libgnuradio-wavelet3.7.5 - gnuradio
wavelet functions
libgnuradio-wxgui3.7.5 - gnuradio
wxgui functions
libgnuradio-zeromq3.7.5 - gnuradio
zeromq functions
libvolk-bin - vector optimized
runtime tools
libvolk-dev - gnuradio vector
optimized function headers
libvolk0.0.0 - gnuradio vector
optimized functions
```

Now, install gnuradio first:

```
sudo apt-get install gnuradio
```

Then install gr-osmosdr:

```
sudo apt-get install gr-osmosdr
```

Finally, install the gr-air-modes:

```
sudo apt-get install gr-air-modes
```

Amateur Satellite Tracking

Satellite tracking is definitely one of the simplest fun projects you can do with your Hamshack Raspberry Pi. In order for you to have fun tracking satellites, you will have to have the GPredict software installed first.

GPredict or Gnome Predict, is a free and real-time satellite tracking program. This is an absolutely amazing software that includes these incredible features:

- The tracking number of satellites is unlimited. Well, it is limited but only by your computer's processing power.
- You can display the data in maps, lists, polar plots, or even combine them all together.

- You can have multiple open modules at the same time.
- You can also use many ground stations.
- The information about the real time data, as well as the predicted passes, is extremely detailed.
- Via Hamlib rigctld, it actually allows doppler radio tunings.

Again, the installation process uses the CLI.

First, you need to update your Pi's repository index:

```
sudo apt-get update
```

Then, you need to search the index for GPredict programs:

```
apt-cache search gpredict
```

Now, install the software:

```
sudo apt-get install gpredict
```

To open GPredict, go to the top left Application Menu of the PIXEL GUI, select Education, and then GNOME Predict.

If you want to uninstall it, you can do it with this command:

```
sudo apt-get uninstall gpredict
```

Once you have the GPredict installed on your Pi, you can actually track any real-time conceivable satellite from all around the world.

The next step is only to interface your Pi with an antenna controller, and start having fun with tracking real-world satellites.

XLog Logging

Just like your phone, your ham radio also needs to have an ordered log list. Keeping track of the logs is very important. XLog is actually an amateur radio logging program that keeps track of the logs, and places the most recent contacts on top of the list. The logs are kept in a text file, and the QSOs are shown in a list.

This program is super convenient because it actually enables you to add, remove, and even update the list.

There is a dxcc information displayed for every made contact, where the distance and bearing are calculated, whether it is a long or a short path.

The installation of XLog is done through the PIXEL GUI of your Desktop:

- Open the top left Application Menu from the PIXEL GUI, hit Preferences, and then click on Add/Remove Software.

- Search for the XLog in the window that has appeared.

- Check both of the XLog boxes in order to select them.

- To install XLOG, simply click on the Apply button.

To launch the XLog program, click on the top left Application Menu in the PIXEL GUI, choose Applications, and then click on XLog.

Graphical Transceiver Control Program

The best program for controlling the graphical ham radio transceiver, is definitely Flrig. This program can be used as an adjunct to the previously discussed Fldigi program, but it also works great on its own.

The installation of Flrig includes the Desktop PIXEL GUI:

- Choose the top left Application Menu in your PIXEL GUI, hit Preferences, and then click on Add/Remove Software.

- When the window appears try to find the Flrig program,

- Select the program by clicking and checking its box.

- Install Flrig, by hitting the Apply button.

You can start the program by choosing the top left Application Menu in the GUI, and then clicking on Flrig.

Required Hardware Configuration:

- Raspberry Pi P3 Model B
- Yaesu transceiver, FT-857D
- Yaesu CAT Cable, CT-62
- USB Dual Serial Adapter (Micro Connectors. Inc)

Pi's Ports Configuration:

The configuration is the same just like for the Fldgi program:

- Make sure that the Raspberry Pi P3 Model B is powered on.
- With a powered USB Hub, plug the USB Dual Serial Adapter into the USB port of the Pi
- This is what should appear in the */dev/* directory:
  `dev/ttyUSB0`
  `/dev/ttyUSB1`
- Plug the Yeasu CT-62 Cat Cable into the FT-857D transceiver
- Turn on the FT-857D transceiver
- Connect the Yaesu CT-62 cable to the Serial Dual Adapter RS-232 S1. The RS-232 S1 connector is `/dev/ttyUSB0`.

Flrig Configuration:

- Choose *Config, Setup,* and then *Transceiver*
- Make sure that at the primary tab:
  - The Rig is FT-857D
  - The Ser.Port is `/dev/ttyUSB0`
  - The Baud is 4800
  - The PTT via CAT is checked
  - 2 StopBids
- Finally, push the Init Button

Now, the Flrig program is in control of the FT-857D transceiver.

Morse Code Virtual Radio

Invented in 1836 by Samuel Morse, the Morse code represents the method of sending, as well as receiving messages with the help of beeps (long and short), where the short beep is called a dot, and the long one a dash. This method is what the ham radio users use to communicate to each other, therefore, it is needless to say how being able to decode such a message is important for you.

And this tutorial will show you just that – how to connect a Morse key to the GPIO pins of your Raspberry Pi, as well as how to write and decode these keys. So let's get started.

Playing a Beep Test

Before all, boot your Pi and log in. Since this exercise produces sound, it is recommended to use headphones, not only to avoid distracting other people, but also to keep yourself focused on the exercises.

Redirect to the socket of your headphones:

```
sudo amixer cset numid=3 1
```

In order to make the tone sound, we need some code. It is best to use Python 3 for this. Now start editing a blank file with this command:

```
nano morse-code.py
```

Paste the key below:

```
#!/usr/bin/python3
import pygame
import time
from array import array
from pygame.locals import *

pygame.mixer.pre_init(44100, -16, 1, 1024)
pygame.init()
```

```python
class ToneSound(pygame.mixer.Sound):
    def __init__(self, frequency, volume):
        self.frequency = frequency

pygame.mixer.Sound.__init__(self, self.build_samples())
        self.set_volume(volume)

    def build_samples(self):
        period = int(round(pygame.mixer.get_init()[0] / self.frequency))
        samples = array("h", [0] * period)
        amplitude = 2 ** (abs(pygame.mixer.get_init()[1]) - 1) - 1
        for time in range(period):
            if time < period / 2:
                samples[time] = amplitude
            else:
                samples[time] = -amplitude
        return samples
```

The usual tones of the Morse codes are between 400 and 1000 Hz, and for this exercises, we will use the 800 Hz frequency.

Add this to the bottom of the file:

```
tone_obj = ToneSound(frequency = 800,
volume = .5)

tone_obj.play(-1)   #the -1 means to
loop the sound
time.sleep(2)
tone_obj.stop()
```

Here, the `tone_obj` is the object that is created and the `ToneSound` is the blueprint.

To save, you should press *Ctrl* + *o* and then *Enter*, and to quit press *Ctrl* + *X* and then *Enter*.

Make the file executable:

```
chmod +x morse-code.py
```

Run this code to hear a beep that is long two seconds:

```
./morse-code.py
```

Connecting the Morse Key to the GPIO Pins

For this purpose, you will need to do some physical computing. The goal here is to switch the voltage for the GPIO pin, on and off, to enable changing the reading of the pin whenever we press the key.

A Pull-Up Circuit:

Through a large 10k Ω resistor, wire your GPIO pin to 3.3 volts, to make sure that it will read HIGH. To make the pin go LOW when we press it, we need to short it to the ground with the Morse key.

A Pull-Down Circuit
Here the GPIO pin is wired through the 10k Ω resistor in a way that it always reads LOW. Then, we short it to 3.3 colts through the Morse key to ensure that it will go HIGH when pressed.

It's important to know what both of these methods will do, as they both work just fine. Choosing which one to go with is often just a personal preference.

Now, take two jump wires, and screw their male ends to the Morse Code key's terminal blocks. Choose either the pull up or pull down configuration, and then connect the female ends to the corresponding GPIO Pins of your Pi.

For this example, we are using the pull up option. You can choose the pull down configuration instead, just remember that in that case the code has to be altered.

Detecting the Key Position

Edit the previous tone program by entering this command:

```
nano morse-code.py
```

In order for you to access the GPIO pins, you will have to import the gpiozero library:

```
import pygame
import time
from RPi import GPIO
```

Make sure to remove these next lines from the bottom. We will put them back later:

```
tone_obj.play(-1)
time.sleep(2)
tone_obj.stop()
```

Now, copy the text below:

```
pin = 4
key = gpio.Button(pin, pull_up=True)
```

```
while True:
    reading = key.is_pressed
    print("ON" if reading else "OFF")
    time.sleep(1)
```

Then, press *Ctrl +o* and *Enter* to save, and *Ctrl + x* and *Enter* to quit.

If you too have chosen the pull up configuration, when the key is up it should read OFF. Hold it for a couple pf seconds until it reads ON.

```
OFF
OFF
OFF
ON
ON
ON
OFF
OFF
OFF
```

Quit by pressing *Ctrl + C*.

Playing a Tone

Only if the program responds by playing and stopping each time that we press the key, the Morse Code Virtual Radio can actually work.

Take a look at the following code:

```
tone_obj = ToneSound(frequency = 800, volume = .5)

pin = 4
key = gpio.Button(pin, pull_up=True)

print("Ready")

while True:
    key.wait_for_press()
    tone_obj.play(-1)   #the -1 means to loop the sound
    key.wait_for_release()
    tone_obj.stop()
```

To edit, enter this command:

```
nano morse-code.py
```

The ToneSound should be at the top of your program and the while loop from before should be deleted. Enter the code from above now.

To test your code, enter:

```
./morse-code.py
```

You can quit by pressing *Ctrl + C*.

Decoding the Morse

In order to be able to decode the More Codes, you need to know the difference between a dot and a dash. In general, a dot lasts for about 0.15 seconds. If it is longer than that, then chances are it is a dash.

Here is a simple code:

```
tone_obj = ToneSound(frequency = 800, volume = .5)

pin = 4
key = gpio.Button(pin, pull_up=True)

DOT = "."
DASH = "-"

key_down_time = 0
key_down_length = 0

print("Ready")

while True:
    key.wait_for_press()
    key_down_time = time.time()
#record the time when the key went down
```

```
    tone_obj.play(-1) #the -1 means
to loop the sound
    key.wait_for_release()
    key_down_length = key_up_time -
key_down_time #get the length of time
it was held down for
    tone_obj.stop()

    if key_down_length > 0.15:
        print(DASH)
    else:
        print(DOT)
```

To edit, enter:

```
nano morse-code.py
```

You can either save or quit at this point. Test the code with:

```
./morse-code.py
```

Your output should look like this:

.

.

.

-

-

.

.

.

Translating the Code into Text

The main point obviosuly is to be able to translate the code to text in order to communicate.

This code will help you find the corresponding numbers and letters for the dots and dashes:

```
wget https://goo.gl/aRju1j -O morse_lookup.py --no-check-certificate
```

To edit, enter this:

```
nano morse_lookup.py
```

The morse_code_lookup serves as a dictionary object. With it, you can actually create your own dictionary and translate between two languages.

```
english_to_french = {
    "Hello": "Bonjour",
    "Yes": "Oui",
    "No": "Non"
}

print(english_to_french["Hello"])
```

Here, the result would be `Bonjour`.

Adding the Code

Again, to edit, enter this:

```
nano morse_lookup.py
```

Now, we add two variables. The first one will record the time of the key release in order to measure the silent gaps. The second one holds the list of dashes and dots.

```
key_up_time = 0
buffer = []
```

Now, let's add the variables to this code:

```
key_down_time = 0
key_down_length = 0
key_up_time = 0
buffer = []

print("Ready")

while True:
    key.wait_for_press()
    key_down_time = time.time() #record the time when the key went down
    tone_obj.play(-1) #the -1 means to loop the sound
```

```
    key.wait_for_release()
    key_up_time = time.time() #record the time when the key was released
    key_down_length = key_up_time - key_down_time #get the length of time it was held down for
    tone_obj.stop()
    buffer.append(DASH         if key_down_length > 0.15 else DOT)
```

Add *Crtl +o* and then *Enter* to save.
Now, add this:

```
#!/usr/bin/python3
import pygame
import time
import gpiozero as gpio
import _thread as thread
from array import array
from pygame.locals import *
from morse_lookup import *
```

Then, you need to put in the code that will run on an entirely different thread:

```
def decoder_thread():
    global key_up_time
    global buffer
    new_word = False
    while True:
```

```
        time.sleep(.01)
        key_up_length = time.time() - key_up_time
        if len(buffer) > 0 and key_up_length >= 1.5:
            new_word = True
            bit_string = "".join(buffer)
            try_decode(bit_string)
            del buffer[:]
        elif new_word and key_up_length >= 4.5:
            new_word = False
            sys.stdout.write(" ")
            sys.stdout.flush()
```

The final code should look like this:

```
#!/usr/bin/python3
import pygame
import time
import gpiozero as gpio
import _thread as thread
from array import array
from pygame.locals import *
from morse_lookup import *

pygame.mixer.pre_init(44100, -16, 1, 1024)
pygame.init()
```

```python
class ToneSound(pygame.mixer.Sound):
    def __init__(self, frequency, volume):
        self.frequency = frequency

        pygame.mixer.Sound.__init__(self, self.build_samples())
        self.set_volume(volume)

    def build_samples(self):
        period = int(round(pygame.mixer.get_init()[0] / self.frequency))
        samples = array("h", [0] * period)
        amplitude = 2 ** (abs(pygame.mixer.get_init()[1]) - 1) - 1
        for time in range(period):
            if time < period / 2:
                samples[time] = amplitude
            else:
                samples[time] = -amplitude
        return samples

def decoder_thread():
    global key_up_time
    global buffer
```

```python
    new_word = False
    while True:
        time.sleep(.01)
        key_up_length = time.time() - key_up_time
        if len(buffer) > 0 and key_up_length >= 1.5:
            new_word = True
            bit_string = "".join(buffer)
            try_decode(bit_string)
            del buffer[:]
        elif new_word and key_up_length >= 4.5:
            new_word = False
            sys.stdout.write(" ")
            sys.stdout.flush()

tone_obj = ToneSound(frequency = 800, volume = .5)

pin = 4
key = gpio.Button(pin, pull_up=True)

DOT = "."
DASH = "-"

key_down_time = 0
key_down_length = 0
key_up_time = 0
```

```
buffer = []

thread.start_new_thread(decoder_thread, ())

print("Ready")

while True:
    key.wait_for_press()
    key_down_time = time.time() #record the time when the key went down
    tone_obj.play(-1)  #the -1 means to loop the sound
    key.wait_for_release()
    key_up_time = time.time() #record the time when the key was released
    key_down_length = key_up_time - key_down_time #get the length of time it was held down for
    tone_obj.stop()
    buffer.append(DASH if key_down_length > 0.15 else DOT)
```

Conclusion

What are you waiting for? This is such an inexpensive and truly fun hobby that can be used to introduce software science to anybody while having a great time. Turn on your own ham radio, and let's explore the world from a high frequency.
Thank you for reading this book and have a great time playing over the air!

Did you find this book helpful? Please, leave a review and let others know about it. Your feedback will be greatly appreciated.

www.ingramcontent.com/pod-product-compliance
Lightning Source LLC
Chambersburg PA
CBHW070309230526
45470CB00002B/796